Graphic Novel Series On Fractions
Book - 2

Keeping it Simple!

Experiencing Simple Fractions and Equivalent Fractions In The Real World

CuriousDots

Many thanks to the children, parents and teachers for adopting and experiencing the **SAPID learning experience**™ and providing valuable feedback.

Print Edition ISBN-13: 978-0-9893751-9-1

Dear Reader,

Welcome to a graphic and fun way of learning mathematics! The **SAPID learning experience**™ was created by a mathematician and a mother, when she was looking for a fun way to introduce pre-algebra topics, such as fractions and decimals, to her daughter. Though there are many approaches and books that teach mathematics, the goal seems to be to get a child ahead of the topics and levels allocated for their age. Children go through the grind of doing pages and pages of the same concept over and over again. While for anyone, let alone children, this gets boring and frustrating, it has the danger of a child developing an aversion to mathematics forever. We see many adults with a mindset that mathematics is not their cup of tea.

Our vision is to make learning fun! And when one has fun, there are no boundaries or levels or age. How far one can go is unlimited!   Our mission is to nurture and expand the creativity in a child. Not to set boundaries through structured approaches that diminish their fun and creativity.

The **SAPID learning experience**™ first introduces a concept through a story using a graphic novel format. This is followed by more stories to understand various aspects of a concept. The stories are short, mostly built around animal characters and are interactive, weaving practice and experiencing of concepts as the story unfolds. It enhances children's creativity and innovation by engaging them in art, lets them put on their detective hat and do some investigation and share their thinking on various solutions to a given situation.

When you are learning a concept such as fractions, many related topics are explored. A concept such as a simple fraction cannot happen in isolation in real life. Other topics such as graphs or measurements or geometry are intertwined naturally and together the stories are shaped into everyday scenarios, taking place in a kitchen or at a game or a playdate. At the heart of these books is the well-known philosophy that stories stick. Revisiting topics are important and stories are easy to recollect.

While some amount of practice is important, this approach will drastically reduce the need for repeated practice on paper. But practice will happen spontaneously as you weave mathematics into everyday life.

Without feedback from several children, parents and teachers this book would not have happened. Please write to us at curiousdots@gmail.com with feedback and suggestions to make learning fun.

CuriousDots team

CuriOusDots

TABLE OF CONTENTS

FRIDAY IS GAME DAY! ⋯⋯⋯⋯⋯⋯⋯⋯⋯⋯⋯⋯⋯ 1

CLASS BOOKS SORTING ⋯⋯⋯⋯⋯⋯⋯⋯⋯⋯⋯ 11

WHEN SIMPLIFYING IS NO LONGER SIMPLE ⋯⋯⋯ 17

THEY ARE EVERYWHERE ⋯⋯⋯⋯⋯⋯⋯⋯⋯⋯ 25

STORY DISCUSSION ⋯⋯⋯⋯⋯⋯⋯⋯⋯⋯⋯⋯ 27

CREATE YOUR OWN STORY ⋯⋯⋯⋯⋯⋯⋯⋯⋯ 28

How To Use This Book:

There are many ways to approach a book of this kind. One way would be to read the book end-to-end like a self-help manual. While this approach will benefit adults such as parents and teachers, for children this book needs to be used in a collaborative setting such as 1-on-1 interaction with a parent or educator or in a classroom setting.

Parents bond with their children by reading daily, especially during bedtime. This book can be used in a similar way. Open dialog and daily conversation about key concepts will make mathematics as much fun as other activities, such as reading and sports.

Children will benefit from reading each of the stories in a group setting or with a parent or an educator. They will naturally get involved in the stories and practice the exercises along with the story characters. Using the questions in the *STORY DISCUSSION* section, children can talk over each story and explore the topic from various angles.

The section *THEY ARE EVERYWHERE* provides tips to discover simple and equivalent fraction in familiar everyday surroundings.

The section *CREATE YOUR OWN STORY* encourages the children's creativity, as they make up their own story with a set of keywords. If a child wants to get super creative and sketch a graphic novel, we have that covered as well! A few graphic novel templates are provided at the end of the book.

The entire book could be completed in a day or within four to five days.

Enjoy!

IT WAS FRIDAY MORNING, THE SNAKES FAVORITE DAY OF THE WEEK. MR.SALVADOR WAS ABOUT TO INTRODUCE A NEW GAME.

THE BEST PART OF IT WAS, MR. SALVADOR ALWAYS HAD SOME PRIZES TO GIVE AWAY AT THE END OF THE GAME!

MR. SALVADOR TOOK OUT A BOX OF HIGHLIGHTERS.

CuriOusDots

MR. SALVADOR TOOK OUT FIVE HIGHLIGHTERS. THE SNAKES NOTICED THAT EACH HIGHLIGHTER WAS MADE OF MULTIPLE COLORS.

Whoever gets this right will get a highlighter!

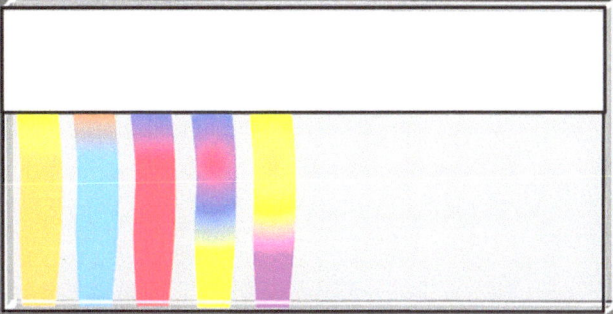
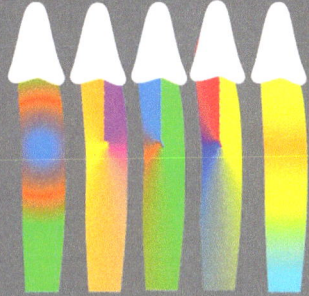

THAT CREATED MORE EXCITEMENT IN THE CLASS!

HISSY EASILY COMPLETED THE FIRST PART OF THE GAME.

Five by ten will represent five highlighters out of the total of ten highlighters.

NEXT, SHE DIVIDED THE RECTANGLE INTO TEN EQUAL PARTS...

AND COLORED FIVE PARTS.

What is the number of highlighters outside the box?
Try writing your answer as a fraction!

$$\frac{5}{10}$$

Color the portion of highlighters outside the box

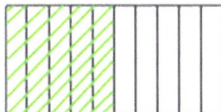

I am confident in my work...

SHE NOTICED THAT GRASSY WHO WAS SITTING NEXT TO HER HAD ALREADY COMPLETED HER WORK.

Grassy's approach is different?

GRASSY HAD WRITTEN ONE BY TWO AND HAD DIVIDED THE RECTANGLE INTO ONLY TWO PARTS.

What is the number of highlighters outside the box? Try writing your answer as a fraction!

$$\frac{1}{2}$$

Color the portion of highlighters outside the box

LOOKING AT THEIR SOLUTIONS SIDE BY SIDE, THE COLORING LOOKED SIMILAR BUT THEIR ANSWERS WERE NOT THE SAME.

HISSY

What is the number of highlighters outside the box? Try writing your answer as a fraction!

$$\frac{5}{10}$$

Color the portion of highlighters outside the box

GRASSY

What is the number of highlighters outside the box? Try writing your answer as a fraction!

$$\frac{1}{2}$$

Color the portion of highlighters outside the box

THEY DECIDED TO WAIT AND SEE WHAT OTHERS HAD DONE.

MR. SALVADOR COLLECTED ALL THE GAME SHEETS AND PINNED THEM ON THE WALL.

THE SOLUTION FELL INTO TWO GROUPS. SOME HAD WRITTEN ONE BY TWO AND SOME HAD WRITTEN FIVE BY TEN!

Friday Is Game Day!

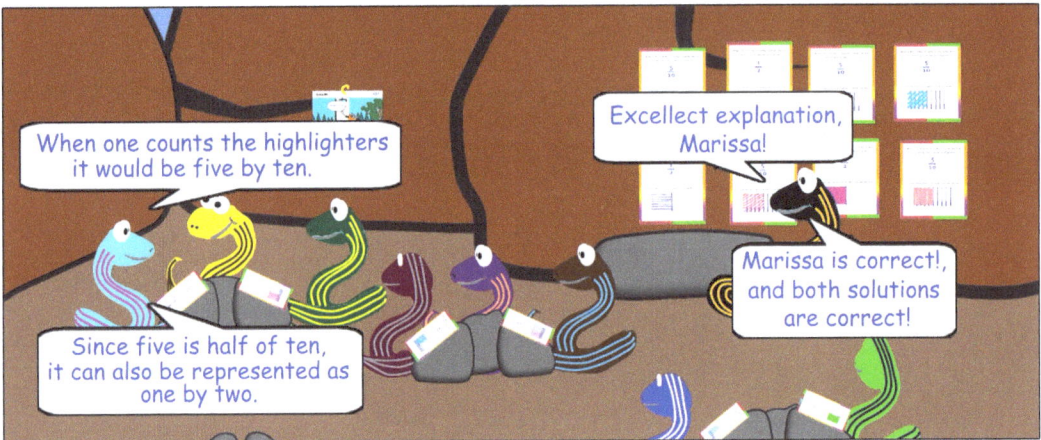

IN THE FIRST SOLUTION, YOU TOOK A GENERAL APPROACH WITHOUT MENTIONING THE NUMBER OF HIGHLIGHTERS.

First approach

Simple Fraction

$$\frac{1}{2}$$

IN THIS APPROACH, ONE BY TWO IS CALLED A **SIMPLE FRACTION!**

IN THE SECOND SOLUTION, YOU COUNTED THE ACTUAL NUMBER OF HIGHLIGHTERS. FIVE BY TEN IS CALLED AN **EQUIVALENT FRACTION** OF ONE BY TWO.

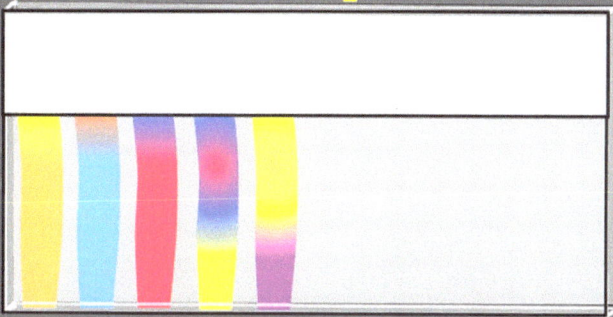
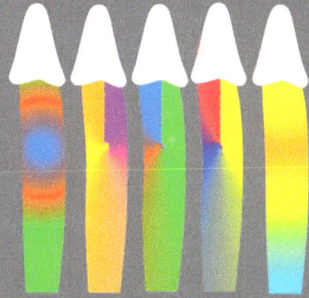

MR. SALVADOR HANDED OUT SOME WORKSHEETS FOR THE CLASS TO EXPLORE CREATING EQUIVALENT FRACTIONS.

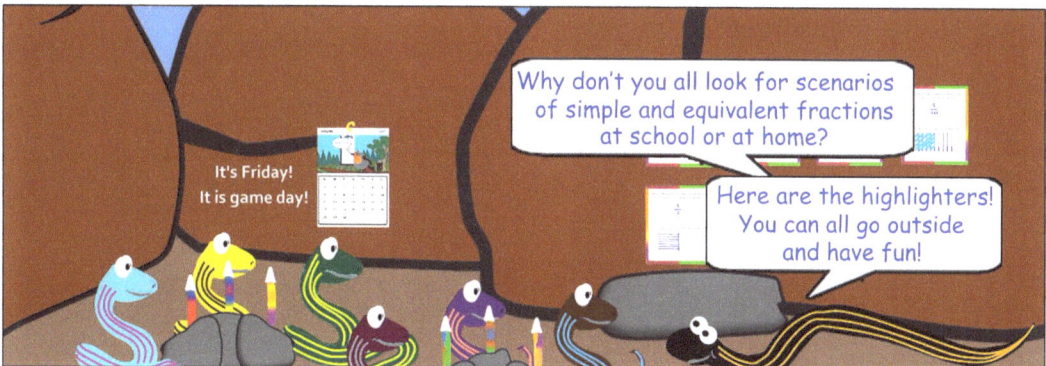

HERE IS THE WORKSHEET THAT MR. SALVADOR HANDED TO THE SNAKES. TRY IT OUT! THE SNAKES SPENT TIME INDIVIDUALLY AND LATER IN GROUPS TO COMPLETE THE WORK SHOWN IN THE NEXT PAGE.

Creating Equivalent Fractions - Whose is the winner?

DRAW A SHAPE AND SPLIT THE SHAPE INTO FOUR EQUAL PARTS. COLOR THREE-FORTH OF THE PARTS.

NOW SPLIT EACH OF THE ABOVE PARTS INTO THREE EQUAL PARTS. IS THERE ANOTHER WAY TO EXPRESS THE SHADED PORTION?

WHY IS 9/12 [OOOPS!] AN EQUIVALENT FRACTION OF 3/4?

WHICH OF THE OPERATORS [$+$, $-$, $\times$, $\div$] CAN HELP CONVERT 3/4 TO 9/12? EXPLAIN YOUR APPROACH.

WHICH SINGLE OPERATOR CAN BE CONSIDERED A WINNER TO EASILY CREATE AN EQUIVALENT FRACTION?

Creating Equivalent Fractions - Whose is the winner?

DRAW A SHAPE AND SPLIT THE SHAPE INTO FOUR EQUAL PARTS.
COLOR THREE-FORTH OF THE PARTS.

$\frac{3}{4}$ is shaded

NOW SPLIT EACH OF THE ABOVE PARTS INTO THREE EQUAL PARTS. IS
THERE ANOTHER WAY TO EXPRESS THE SHADED PORTION?

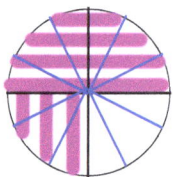

$\frac{9}{12}$ is shaded

WHY IS 9/12 [OOOPS!] AN EQUIVALENT FRACTION OF 3/4?

They both represent the same amount.

WHICH OF THE OPERATORS [$+$, $-$, $\times$, $\div$] CAN HELP CONVERT
3/4 TO 9/12? EXPLAIN YOUR APPROACH.

There are two approaches.
1. Using repeated addition based on the number of parts.

$$\frac{3 + 3 + 3}{4 + 4 + 4} = \frac{9}{12}$$

2. Or multiplying by the number of parts.

$$\frac{3 \times 3}{4 \times 3} = \frac{9}{12}$$

WHICH SINGLE OPERATOR CAN BE CONSIDERED A WINNER TO EASILY
CREATE AN EQUIVALENT FRACTION?

Multiplication is the winner! One can multiply the
numerator and denominator by the same number to create
an equivalent fraction.

CuriousDots

Class books sorting

SUNNY AND RAINBOW WERE EXCITED ABOUT THEIR CLASS JOB THIS WEEK. THEY HAD THE TASK OF SORTING AND CATALOGING THE CLASS BOOKS INTO DIFFERENT CATEGORIES.

"Be awesome! Be a book nut!" —Dr. Seuss

We have twenty chapter books!

We have ten picture books and ten biography books.

"Be awesome! Be a book nut!" —Dr. Seuss

We have to give this information to our school administrator, Mr. Smarts.

Yes, I heard that Mr. Smarts is creating an online catalog of all books we have in school!

Let us try to make the information colorful and nice!

RAINBOW TYPED THE DETAILS ABOUT THEIR CLASS BOOKS AND FELT GOOD ABOUT HER WORK.

Mr. Salvador's Class Book Catalog

Biography Books - 10
Picture Books - 10
Chapter Books - 20

How does this look?

It looks great!

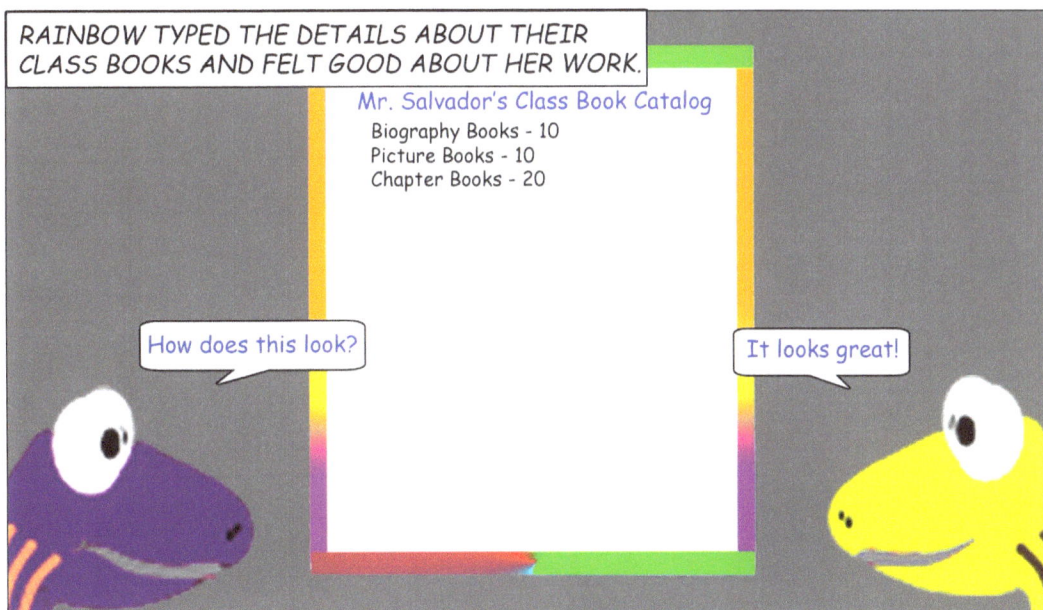

A Bar Graph (also called Bar Chart) is a graphical display of data using bars of different heights.

Imagine you did a survey of your friends to find which fruit they liked best:

Table: Favorite Fruit			
Apple	Orange	Banana	Strawberries
3	4	1	5

We can show the data on a bar graph like this:

Should we include a bar chart on the category of books?

That will make the presentation colorful!

I think that will be interesting!

My mom always says that a picture is worth a thousand words!

SUNNY DREW A BAR CHART. HE STARTED NUMBERING THE BOOKS ON THE GRAPH AND REALISED THAT 40 NUMBERS WOULD BE A LOT TO WRITE! THEY DECIDED TO REDUCE THE VALUES USING SIMPLE FRACTIONS.

Total Class Books - 40	Fraction of books
Biography Books - 10	$\frac{10}{40}$
Picture Books - 10	$\frac{10}{40}$
Chapter Books - 20	$\frac{20}{40}$

Hmmm.... Can we reduce the values using simple fractions?

That might make it easier and we may not need a large sheet?

You need a larger sheet to write forty numbers!

RAINBOW FIRST REPRESENTED THE CATEGORIES OF BOOKS AS FRACTIONS.

Great idea!

1/4 is the simple fraction for 10/40 because you can divide both 10 and 40 with the common number 10. And 1/4 cannot be reduced further.

DO YOU AGREE WITH SUNNY'S EXPLANATION ABOUT SIMPLE FRACTION?

CuriousDots

CAN YOU HELP RAINBOW AND SUNNY COMPLETE THE FRACTIONS AND THE BAR CHART BELOW?

Mr. Salvador's Class Book Catalog

Total Class Books - 40	Fraction of books	Simple fraction
Biography Books - 10	$\frac{10}{40}$	$\frac{1}{4}$
Picture Books - 10	$\frac{10}{40}$	
Chapter Books - 20	$\frac{20}{40}$	

Bar Chart

$$\frac{1}{4}$$

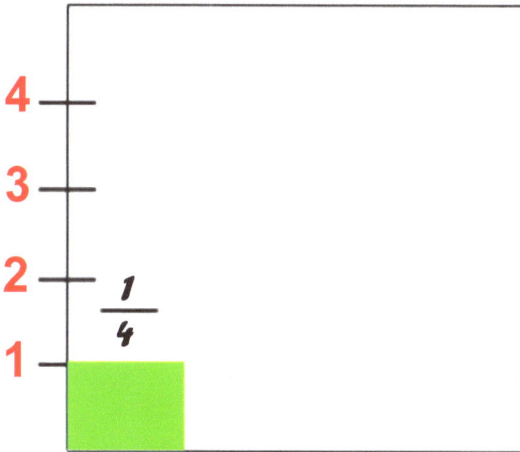

Biography Books   Picture Books   Chapter Books

Class Books Sorting

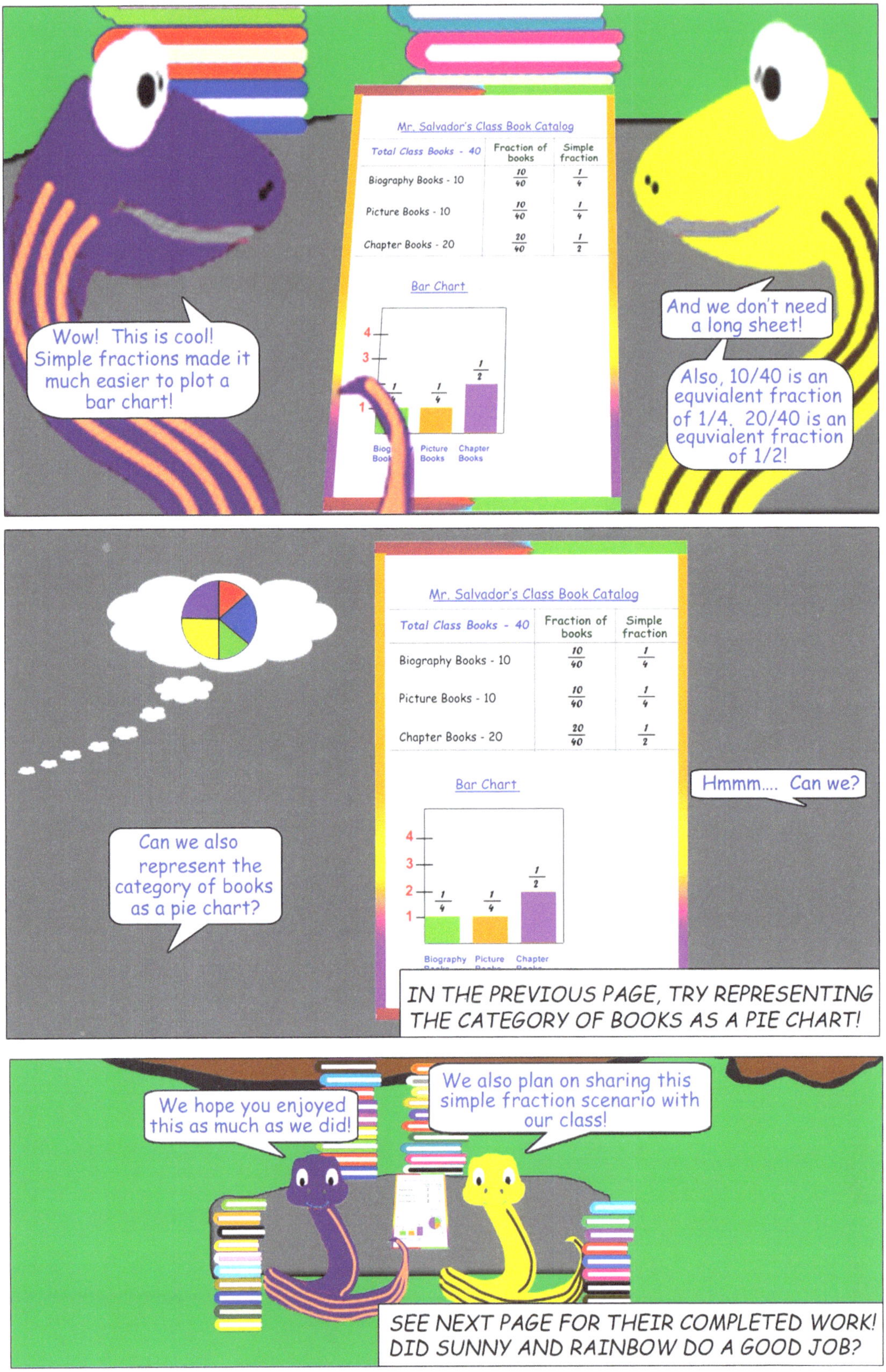

Mr. Salvador's Class Book Catalog

Total Class Books - 40	Fraction of books	Simple fraction
Biography Books - 10	$\frac{10}{40}$	$\frac{1}{4}$
Picture Books - 10	$\frac{10}{40}$	$\frac{1}{4}$
Chapter Books - 20	$\frac{20}{40}$	$\frac{1}{2}$

Bar Chart

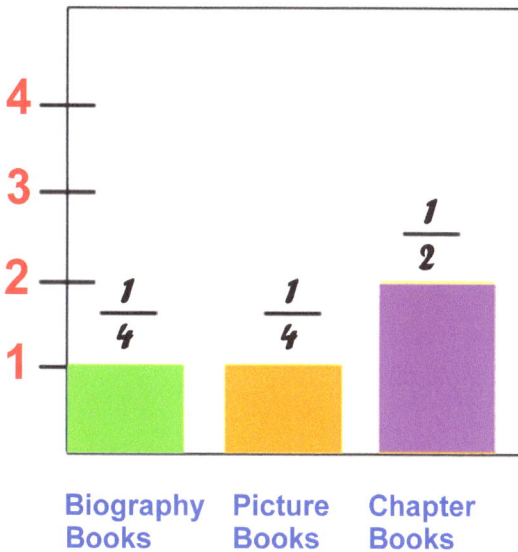

Pie Chart

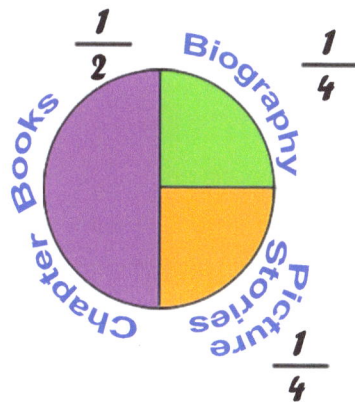

Page Intentionally Left Blank

When Simplifying is no longer simple

THE PIGS HAD JUST FINISHED DELICIOUS CUPCAKES AT THEIR LUNCH. MR SPICE, ONE OF THEIR FAVORITE CHEFS AT SCHOOL, WAS RETIRING TODAY.

Thank you all for the wonderful painting.

I will greatly miss this place.

I made some additional cupcakes for the neighborhood shelter. Can you all help Ms. Candy pack them?

YES Mr. Spice!

MS. CANDY, ANOTHER CHEF, ROLLED IN A CART FULL OF CUPCAKES AND EMPTY BOXES OF VARIOUS SIZES.

Thanks for helping with the packing of these cupcakes.

We have 24 peanut jelly cupcakes. We also have 30 cupcakes, some of which are lemon and some are chocolate.

I will write the rules on the board, Ms. Candy!

There are some rules to follow in packing these.

If you can dream it, you can do it. - *Walt Disney*

Rules for packing

1. The peanut jelly cupcakes will need to be packed separately.
2. Use the same size boxes for packing all the cupcakes.
3. Use the minimum number of boxes.

Total cupcakes - 54
Peanut jelly cupcakes - 24
Other cupcakes - 30

AS MS. CANDY DESCRIBED THE RULES, BANKO WROTE THEM ON THE BOARD.

Rules for packing

1. The peanut jelly cupcakes will need to be packed separately.
2. Use the same size boxes for packing all the cupcakes.
3. Use the minimum number of boxes.

Total cupcakes - 54
Peanut jelly cupcakes - 24
Other cupcakes - 30

We cannot put the 24 peanut jelly cupcakes in a box and the other 30 in another box.

How about putting each of the cupcake in a single box?

Then, how about putting 2 cupcakes in a box? That would be 27 boxes.

That would need a total of 54 boxes! Is that the best option?

Actually, I think we can do 3 cupcakes in a box!

Yes, then the box sizes will be different. That will break rule 2!

Rule 3 says to use the minimum number of boxes?

THEY DISCUSSED HOW TO SPLIT THE CUPCAKES WITHOUT BREAKING THE RULES.

If you can dream it, you can do it. - *Walt Disney*

Rules for packing

1. The peanut jelly cupcakes will need to be packed separately.
2. Use the same size boxes for packing all the cupcakes.
3. Use the minimum number of boxes.

Total cupcakes - 54
Peanut jelly cupcakes - 24
Other cupcakes - 30

Great ideas everyone! Why don't you write your discussions on the board?

CuriousDots

BANKO CAPTURED THEIR DISCUSSION AND *lt Disney*
ASKED IF THERE WERE ANY MORE IDEAS.

1. The peanut jelly cupcakes will need to be packed separately.
2. Use the same size boxes for packing all the cupcakes.
3. Use the minimum number of boxes.

Total cupcakes - 54
Peanut jelly cupcakes - 24
Other cupcakes - 30

Number of cupcakes in one box	1 cupcake	2 cupcakes	3 cupcakes
Number of boxes for peanut jelly cupcakes	24	12	8
Number of boxes for other cupcakes	30	15	10

I think we should continue with 5, 6, 7 cupcakes and so on until we find the size that will fit the maximum cupcakes for all of them.

THE CLASS CALCULATED UPTO 15 CUPCAKES IN A BOX.
THEY ALL OBSERVED THE INFORMATION QUIETLY.

1. The peanut jelly cupcakes will need to be packed separately.
2. Use the same size boxes for packing all the cupcakes.
3. Use the minimum number of boxes.

Total cupcakes - 54
Peanut jelly cupcakes - 24
Other cupcakes - 30

Number of cupcakes in one box	1 cupcake	2 cupcakes	3 cupcakes	4 cupcakes	5 cupcakes	6 cupcakes
Number of boxes for peanut jelly cupcakes	24	12	8	6		4
Number of boxes for other cupcakes	30	15	10		6	5

Number of cupcakes in one box	8 cupcake	10 cupcakes	12 cupcakes	15 cupcakes
Number of boxes for peanut jelly cupcakes	3		2	
Number of boxes for other cupcakes		3		2

If you can dream it, you can do it. - Walt Disney

I agree!

Then we will not have the same size boxes for all the cupcakes.

We should not consider numbers like 4 or 5 cupcakes, as they cannot divide both 24 and 30.

When Simplifying Is No Longer Simple

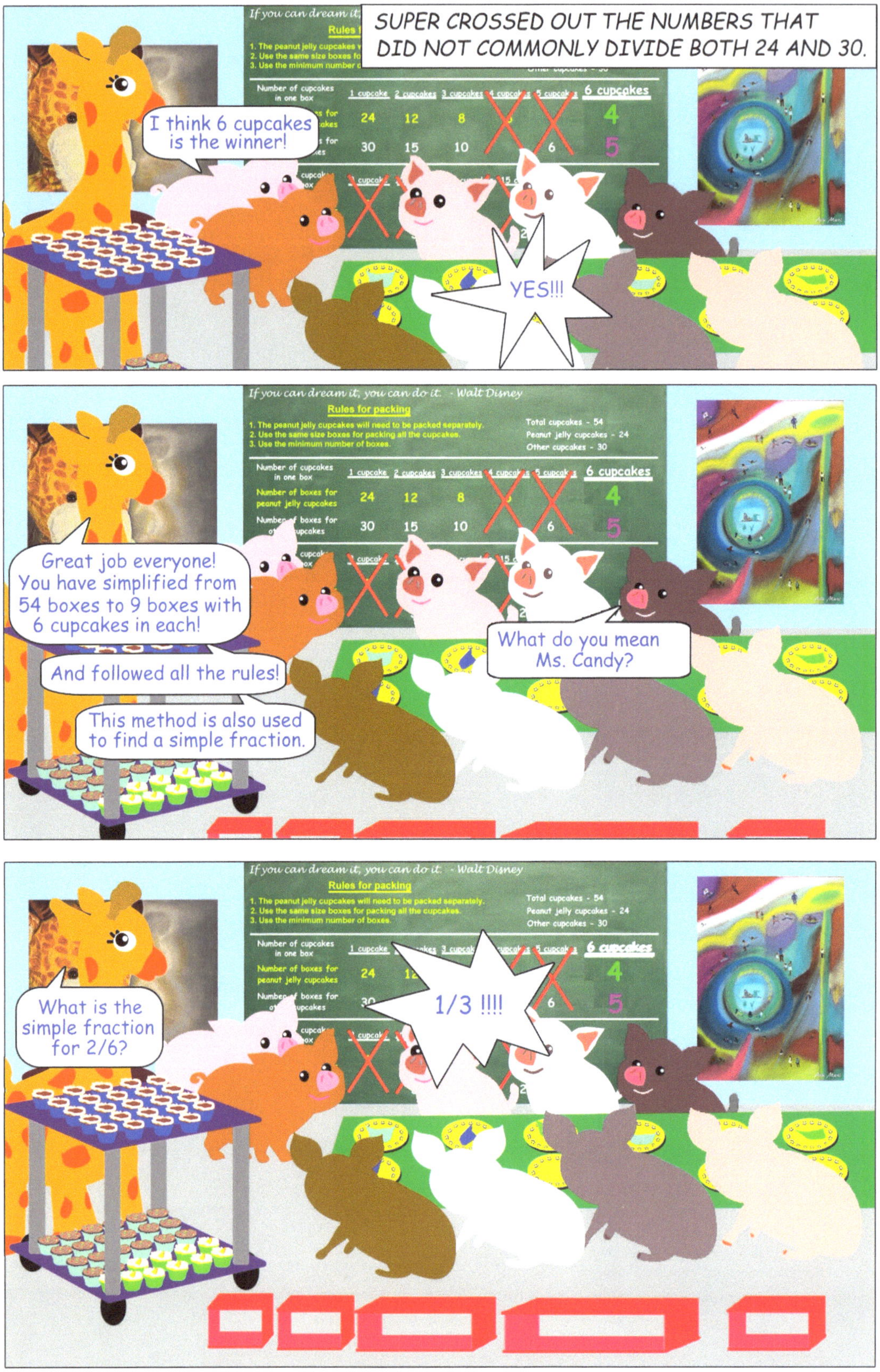

When Simplifying Is No Longer Simple

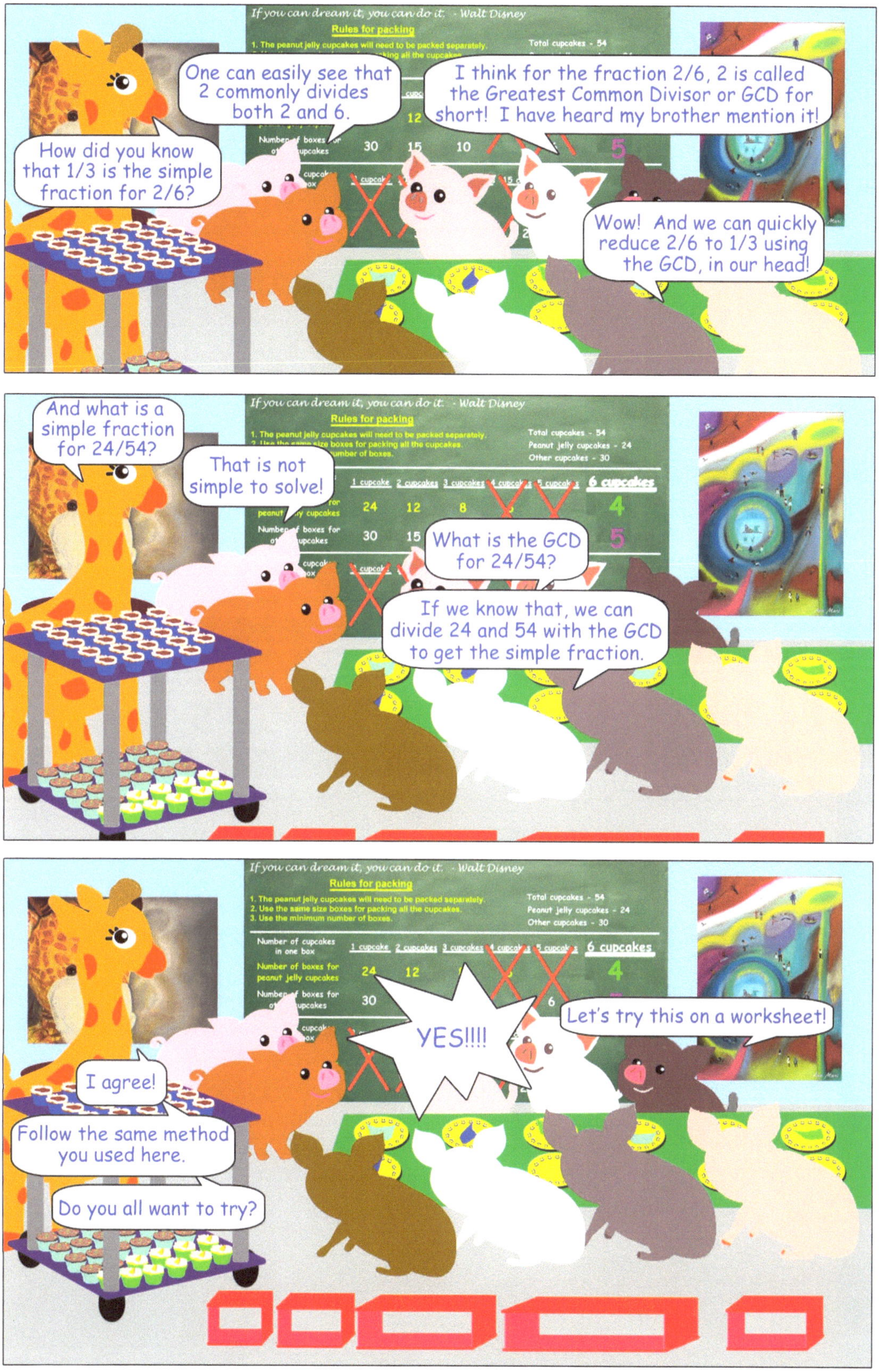

TRY FINDING THE SIMPLE FRACTION FOR 24/54 ALONG WITH THE PIGS.

USE THE SPACE ON THE NEXT PAGE TO FIND THE SIMPLE FRACTION FOR 24/54. ALSO, CHECK THE PIG'S WORK SHOWN AFTER THE NEXT PAGE.

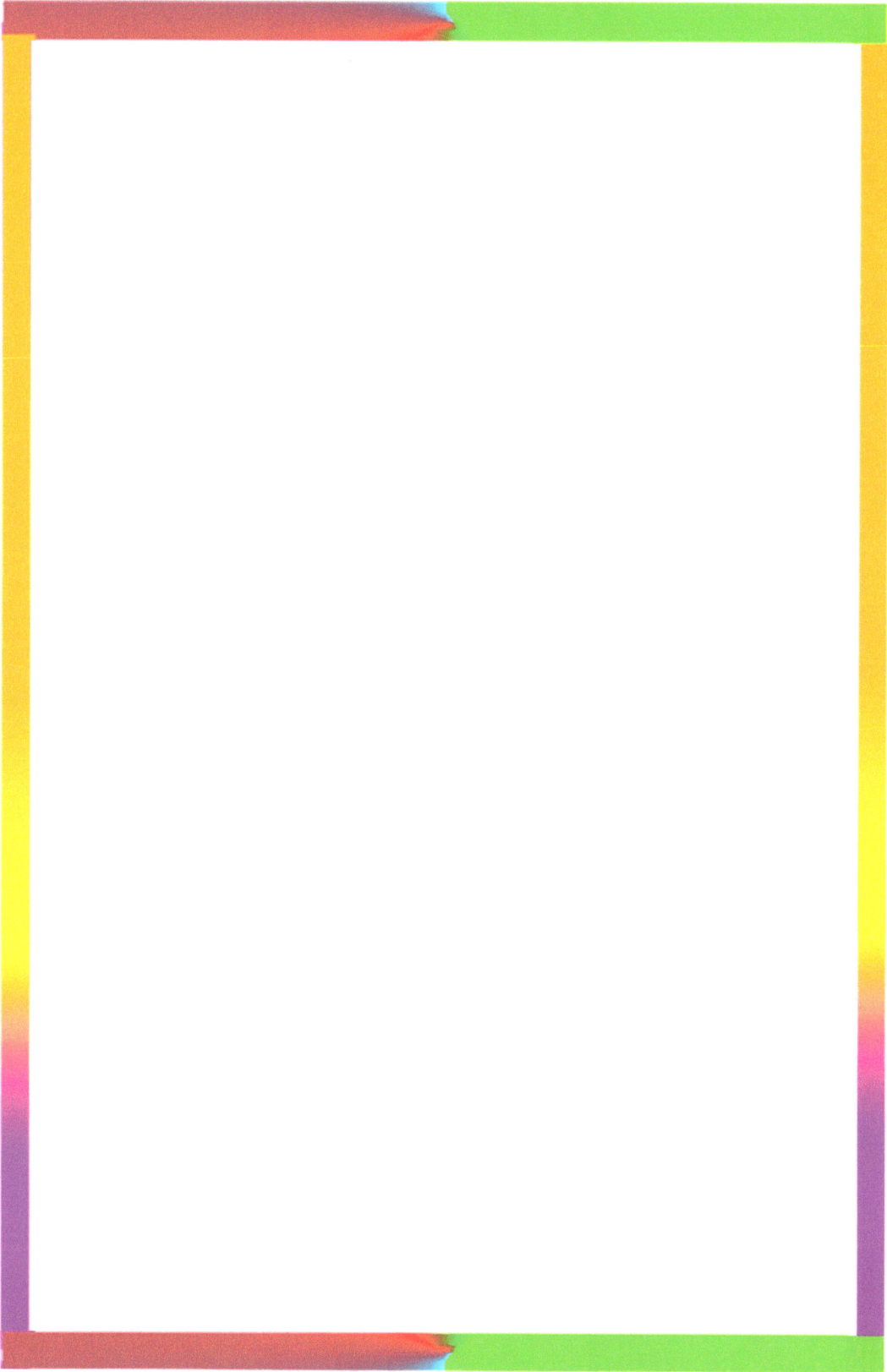

HERE IS THE WORKSHEET THE PIGS USED TO FIND THE SIMPLE FRACTION.

Finding simple fraction for 24/54

Divisors for 24 - 1,2,3,4,6,8,12,24

Divisors for 54 - 1,2,3,6,9,18,27,54

Common Divisors - 1,2,3,6

Greatest Common Divisor - 6

Reducing Numerator - 24 ÷ 6 = 4

Reducing Denominator - 54 ÷ 6 = 9

Simple Fraction of 24/54 - $\dfrac{4}{9}$

CuriousDots

They are everywhere...

Exploring Everyday Examples

Look for items that can be grouped into various categories. Find the fraction for a particular category. Investigate whether it is a simple fraction or an equivalent fraction. If it is an equivalent fraction, convert to its simple fraction. See whether the simple fraction makes understanding the quantity of a particular category easy. Below are few examples, to get you started.

1. Spoons in silverware
2. Particular shaped pan
3. When cutting or dividing any food item
4. A particular fruit or vegetable

Investigation

Simple and equivalent fractions play a key role in data collections and analysis. Below are few examples.

1. Which color t-shirt do you have the most? What is the fraction of this colored t-shirts compared to the total t-shirts you have? Is that your favorite color?
2. Log different activities your class does every day including lunch and recess time for a week. What is the fraction of recess time compared to the total activities?
3. Find each student's favorite subject or activity. Is there one that is the most popular? How popular is it?
4. On a particular day, make a log of food items you eat into two categories. The categories are *healthy* and *not so healthy*. How much fraction of healthy food did you eat compared to the total items?

Can you find an equivalent fraction?

Finding an equivalent fraction out there is not easy. In fact, it may not be possible. Here are few places we explored.

1. Have you seen a measuring cup with equivalent fractions?
2. Have you seen measuring spoons with equivalent fractions?
3. Have you seen a recipe with equivalent fractions?
4. Did you ever see an equivalent fraction while driving on a highway or even on a local road?
5. Have you come across equivalent fractions in a user manual for a board game or any other games?

Discuss stories using the following questions as guidelines.

1. What is the predominant theme in the story? Around who is the story centered?
2. Where does the primary action take place?
3. How does the story get started? What is the initial incident?
4. Briefly describe the rising action of the story.
5. What is the high point, or climax, of the story?
6. Discuss the falling action or end of the story.
7. Was there a villain in the story? a hero? a dynamic character?
8. Does the story contain a single effect or impression for the reader? If so, what is it?

Look for objects around you that can be grouped into categories. It could be silverware in your kitchen, different colored building blocks or pens and so on.

Use the space below to draft your own story with your own characters. The characters could even be you and your friends! Use the keywords listed below in your story.

If you want to draw and present your story as a graphic novel, use the templates at the end of the book.

You can even try using your script to enact it before a group or your class.

Keywords: Reduce, Simple Fraction, Equivalent Fraction

CuriousDots

Reproduce as many copies as needed

CuriousDots

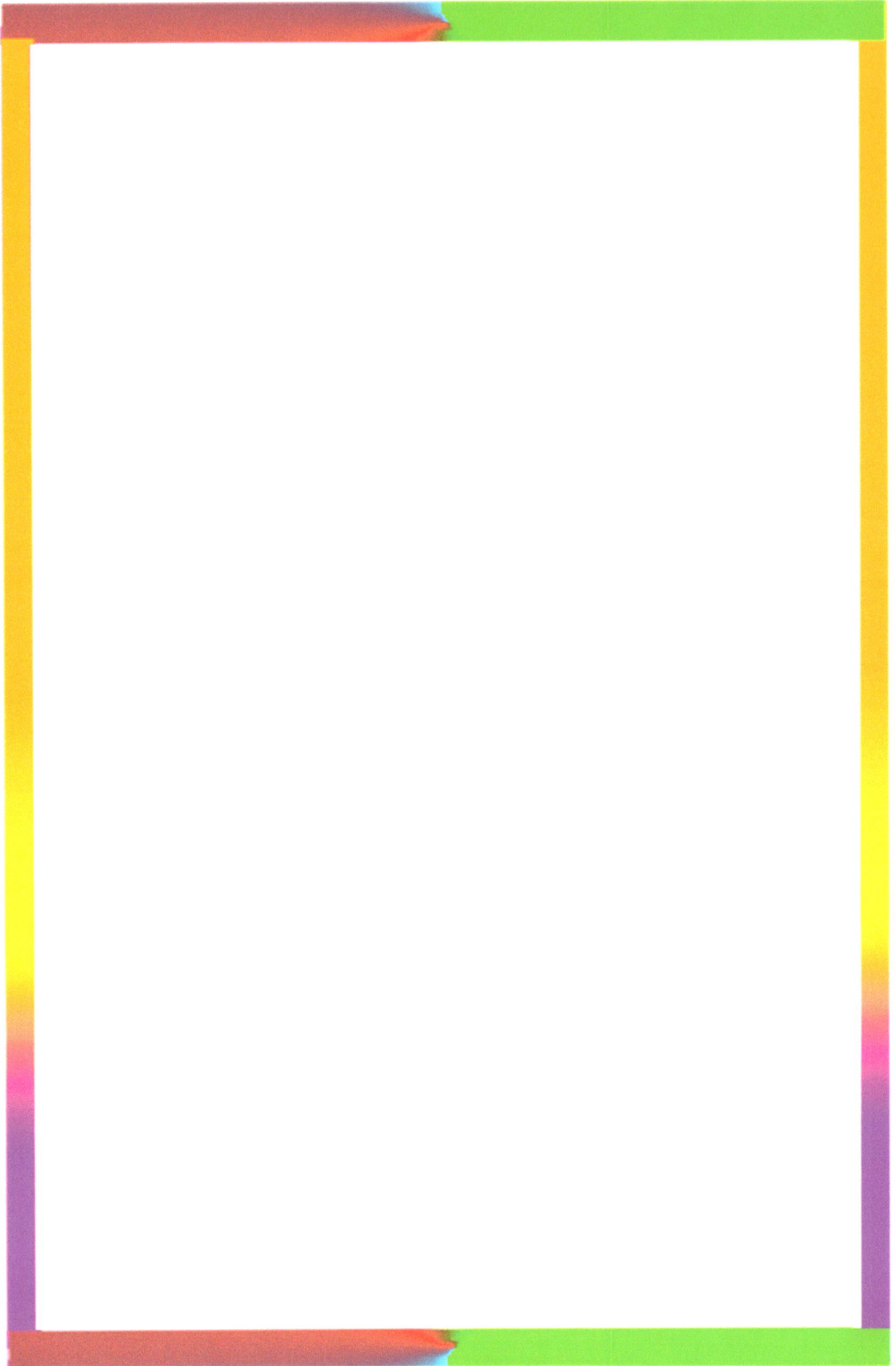

www.ingramcontent.com/pod-product-compliance
Lightning Source LLC
Chambersburg PA
CBHW052044190326
41520CB00002BA/177